THESAURUS
of
ROCK
and
SOIL MECHANICS TERMS

THESAURUS
of
ROCK
and
SOIL MECHANICS TERMS

Compiled by

J P JENKINS and A M SMITH

Rock Mechanics Information Service,
Imperial College of Science and Technology, London

PERGAMON PRESS

OXFORD · NEW YORK · TORONTO · SYDNEY · PARIS · FRANKFURT

U.K.	Pergamon Press Ltd., Headington Hill Hall, Oxford OX3 0BW, England
U.S.A.	Pergamon Press Inc., Maxwell House, Fairview Park, Elmsford, New York 10523, U.S.A.
CANADA	Pergamon Press Canada Ltd., Suite 104, 150 Consumers Road, Willowdale, Ontario M2J 1P9, Canada
AUSTRALIA	Pergamon Press (Aust.) Pty. Ltd., P.O. Box 544, Potts Point, N.S.W. 2011, Australia
FRANCE	Pergamon Press SARL, 24 rue des Ecoles, 75240 Paris, Cedex 05, France
FEDERAL REPUBLIC OF GERMANY	Pergamon Press GmbH, Hammerweg 6, D-6242 Kronberg-Taunus, Federal Republic of Germany

First Edition 1984

Library of Congress Cataloging in Publication Data

Jenkins, J. P.
Thesaurus of rock and soil mechanics terms.
1. Subject headings—Rock mechanics. 2. Subject headings—Soil mechanics. I. Smith, A. M. II. Title.
Z695.1.R63J45 1984 025.4'96241'513 84–6267

British Library Cataloguing in Publication Data

Jenkins, John Philip
Thesaurus of rock and soil mechanics terms.
1. Subject headings—Rock mechanics
2. Subject headings—Soil mechanics
I. Title II. Smith, A.M.
025.4'96241513 Z695.1.R/
ISBN 0-08-031632-8

Printed in Great Britain by A. Wheaton & Co. Ltd., Exeter

INTRODUCTION

The thesaurus was originally compiled by J P Jenkins of the Rock Mechanics Information Service in 1976-1977. Since then it has been used and developed primarily for internal use. With the public availability of the Geomechanics Abstracts data base (GMA) on the Pergamon-InfoLine on-line system, the thesaurus is being published as an aid to users of GMA.

The thesaurus derives in style principally from the Thesaurus of Engineering and Scientific Terms. Other publications consulted include CSIR Rock Mechanics Thesaurus, ASCE Soil Mechanics Thesaurus, US Army Engineer Waterways Experiment Station Microthesaurus of Soil Mechanics Terms, Geodex Soil Mechanics Thesaurus, Canada Mines Branch Mining Thesaurus, and Geosystems' Geosaurus. Details of these publications are given in the Appendix.

Reference has also been made to many textbooks and other publications, and advice sought from colleagues too numerous to mention. This help is gratefully acknowledged.

The thesaurus is a structured list of "controlled" terms (or keywords) used to index the material contained in the Geomechanics Abstracts data base. This data base contains bibliographic references to publications pertaining to the fields of rock and soil mechanics. Coverage includes: properties of rocks and soils, rock and soil masses, geology and hydrogeology, mining, tunnelling, support, foundation engineering, slope stability, comminution, rock and soil improvement techniques, site investigation, laboratory and field tests.

The terms are listed alphabetically with preferred terms in bold type and non-preferred terms in Italic type. For each of the preferred terms, non-trivial relationships with other preferred terms are given, and sometimes a brief scope note or definition is also included. For the non-preferred terms, the user is directed to a preferred term.

The relationships between terms are denoted by the following symbols:

USE	– use (indicates the preferred term which should be used)
UF	– used for (indicates the non-preferred term)
BT	– broader term (indicates a more general term)
NT	– narrower term (indicates a more specific term)
RT	– related term (indicates an associated term which might be considered)

NOTES FOR USERS OF GMA

The controlled terms in the thesaurus appear in the CT= index of GMA; also individual words from the terms will appear in the general IT= index.

eg. "acoustic wave damping" will generate the following index entries:

CT=acoustic wave damping
IT=acoustic
IT=wave
IT=damping

1. In indexing, the narrowest term suitable has been used, and there is no upward posting in GMA. Therefore all narrower terms should be specified in addition to the main term (if required).

   For example, if searching for "carbonate rocks", the narrower terms "chalk", "dolomite", "limestone" and "marble" must be specified for completeness.

   There are two exceptions to this rule:

   i) To permit separation of rocks from soils, the term "rocks" has been added for any item where the words "rock" or "rocks" do not appear in the index terms, for example where an item has been indexed with the name of an individual rock. Soils have been treated similarly.

      Therefore the user may retrieve references on rocks by specifying:

      IT=rock or rocks

      Similarly for soils.

   ii) All tests on rocks and soils have also been indexed with "lab tests" or "field tests".

      Therefore to search for lab tests on rock compressive strength:

      CT=rock compressive strength AND CT=lab tests

2. The indexing terms allocated to GMA items have also been used to produce the annual subject indexes to the printed version Geomechanics Abstracts. In an effort to produce entries more meaningful than single keywords in this printed index where there is no facility for coordination, it has been practice to combine terms together in certain ways. Rules that have been applied in this respect are as follows:

   i) Rock and property may be used eg. rock deformability.

      When searching the CT= index both terms CT=rock deformability and CT=deformability should be used for completeness.

      Similarly for soil and property.

Names of specific rocks may not be combined with a property eg. CT=slates and CT=rock texture should be used, not CT=slate texture.

ii) Any wave term may be combined with any wave phenomena term.

eg. CT=P wave velocity
CT=acoustic wave damping

iii) Design may be combined freely.

eg. CT=earth dam design

3. For specific engineering projects, the PJ=index may be consulted.

eg. PJ=Dinorwic Power Plant

When a project is indexed the term "case studies" together with the name of the country are normally also used.

4. Names of countries have not been included in the thesaurus but are assigned to documents as keywords (UK counties and US states are also added).

Abbreviations used are: UK, USA, USSR.

5. The terms "rock mechanics", "soil mechanics" and "geomechanics" have only been used for general works such as textbooks, or to differentiate rock mechanics from soil mechanics and vice versa.

6. The following terms have been assigned to indicate the type of publication:

bibliographies
case studies
conferences
handbooks
patents
reports
reviews
standards
textbooks
theses

Note: "conferences" is not used to index individual conference papers.

For further details about searching the GMA data base, please refer to the Geomechanics Abstracts Data Base Manual, obtainable from:

Pergamon-InfoLine
12 Vandy Street
London
EC2A 2DE

APPENDIX

1) ENGINEERS JOINT COUNCIL. Thesaurus of Engineering and Scientific terms. New York: Engineers Joint Council, 1964.

2) DOUGLAS, E. Thesaurus of Terms for Information Storage and Retrieval in the Field of Rock Mechanics. 1970. Council for Scientific and Industrial Research report MEG927.

3) Soil Mechanics Thesaurus. J Soil Mech Found Div ASCE, Sept 1967, Part 2, 93(5), 23-147.

4) Geodex Soil Mechanics Thesaurus. In: Geodex Manual and Author Index. USA: Geodex, 1968.

5) ROMANIUK, A S. Thesaurus of Mining Terms. 1969. Canada Department of Energy, Mines and Resources, Mines Branch Information Circular IC225.

6) CHARLES, R (ed). Geosaurus: Geosystems' Thesaurus of Geoscience (3rd ed). London: Geosystems, 1979.

abrasion
RT surface properties.....
wear

abrasiveness
BT surface properties
RT hardness
roughness

absorption
NT gas absorption
moisture absorption
RT attenuation
damping
propagation
waves.....

abutments
RT foundations.....

accelerograms
RT earthquakes
graphical analysis

accelerometers
BT instruments
RT monitoring systems

accidents
RT hazards
safety

accuracy
RT performance
sensitivity

acidic rocks
BT igneous rocks
NT granite
rhyolite

acidity
BT chemical properties

acoustic emission
UF microseismic methods
sound emission
RT noise.....
rock noise

acoustic waves
UF sonic waves
sound waves
RT elastic waves.....
seismic waves.....
ultrasonic waves

acoustics
NT ultrasonics
RT noise.....

additives

adhesion
RT surface properties.....

adits
USE drifts

adsorption
RT surface properties.....
surfaces

aelotropy
USE anisotropy

aeolian soils
NT loess
RT collapsible soils
sands.....
silts

aerial cableways
BT surface structures

aftershocks
RT earthquakes
foreshocks

age
RT Cainozoic rocks/soils
Mesozoic rocks/soils
Palaeozoic rocks/soils
Precambrian rocks/soils

aggregates
UF ballast
BT structural materials
RT gravel
limestone
magnetite
rocks.....
shales
slates

air
RT weathering
wind

air drilling

airborne photography
BT photography
RT photo interpretation
photogrammetry

airborne surveying
BT surveying
RT photogrammetry

aircraft pavements
USE pavements

algorithms
USE data processing

alkaline rocks
USE basic rocks

alloys
BT metals

alluvium

altered rocks
USE metamorphic rocks

analogues
RT electrical methods.....
isopachs

anchorages
PREFER anchors (if appropriate)
RT foundations.....
piers
rock sockets

anchors
UF cable bolts
BT support
RT bolts.....
split set anchors

andesite
BT extrusive rocks
igneous rocks
intermediate rocks

angle of draw
RT subsidence

angular

anhydrite
BT evaporitic rocks
minerals
RT gypsum

anhydrite packing
BT packing
support

anisotropy
UF aelotropy
RT Brazilian tests
indentation tests
isotropy
point load tests

anorthosite
BT minerals
RT feldspar
plagioclase
silicates

anthracite
BT carbonaceous rocks
coal
sedimentary rocks
RT lignite

antiplane strain
RT plane strain

aquifers
RT groundwater
hydrology
wells

arch dams
BT dams
RT dam foundations

arch theory

arches
BT support
RT columns
steel

arching
(load redistribution above an opening)
USE arch theory

arenaceous rocks
BT clastic rocks
sedimentary rocks
RT sands.....
sandstones.....

argillaceous rocks
BT clastic rocks
sedimentary rocks
RT argillites
claystones
marls
mudstone
shales

argillites
BT clastic rocks
metamorphic rocks
sedimentary rocks
RT argillaceous rocks

asperities
RT inclusions
microstructure
roughness
surface properties.....

asphalt
BT structural materials
RT bitumen

attenuation
RT absorption
damping
propagation
waves.....

auger drilling

automation
UF mechanisation

avalanches
RT landslides
rockslides
slip.....

back analysis
analysis of a case study (usually a failure) to calculate properties of the material or structure
BT mathematical analysis

backfill
BT fills
RT cemented fill
earth fill
hydraulic fill

ball mills

ballast
USE aggregates

basalt
BT basic rocks
extrusive rocks
igneous rocks
RT gabbro
volcanoes

base courses
RT foundations.....
pavements
subgrades

basic rocks
UF alkali rocks
ultrabasic rocks
BT igneous rocks
NT basalt
diabase
dunite
gabbro
norite
serpentinite

bauxite
BT minerals
ores

beam tests
BT lab tests
RT bending tests
tension tests.....

beams

bearing capacity
BT mechanical properties
RT punch tests
strength.....

bedding
RT cleavage
fissility
strata

bedding planes
RT strata

bedrock
RT foundations.....

benches
USE berms

bending

bending tests
BT lab tests
RT beam tests
rupture modulus
tensile strength
time dependent behaviour.....

bentonite
BT clays
cohesive soils

bentonite shield
BT shields
tunnelling machines

berms
UF benches
BT surface structures
RT dumps
embankments
slopes.....

biaxial stress

biaxial tests
BT lab tests
RT hollow cylinder tests

bibliographies

Biot theory

bitumen
BT structural materials
RT asphalt

blastability
BT mechanical properties

blasting
NT controlled blasting
presplitting
RT detonation
explosives
fragmentation
loosened zones
rock breaking

block caving
BT caving
mining

blocks
RT interlocking

body waves
BT elastic waves
seismic waves
NT P waves
S waves

bolting machines

bolts
UF rebars
rock/roof bolts
BT support

bolts (contd)
NT grouted bolts
measuring bolts
resin grouted bolts
split set anchors
yielding bolts
RT anchors
pullout tests

borability
BT mechanical properties
RT drillability
penetration rate
tunnelling.....

bord and pillar
BT mining
RT room and pillar

borehole gauges
BT instruments
RT monitoring systems

borehole linings
BT linings

borehole logging
UF well logging
BT field tests
NT crosshole logging
crosshole seismic method

borehole shear tests
BT field tests
shear tests

boreholes
BT holes
RT inclinometers
petroscopes
pressuremeters
pumping tests
shafts.....

boring
NT raise boring
RT mining.....

boring machines
RT tunnelling machines.....

boudinage

boundaries
NT grain boundaries
RT interfaces
mathematical analysis.....

boundary integral method
UF displacement discontinuity method
BT mathematical analysis
numerical analysis

Brazilian tests
indirect tension tests in which cylindrical specimens are diametrally compressed by line loads
UF split cylinder tests
BT lab tests
RT anisotropy
compression tests.....
point load tests
tensile strength
tension tests.....

breakage
USE fragmentation
OR rock breaking

breccia
BT clastic rocks
sedimentary rocks

bridge foundations
BT foundations

bridges
BT surface structures
RT piers

brittle-ductile transition
USE brittle transition

brittle failure
BT failure
RT Griffith criterion

brittle fracture
RT fractures.....

brittle materials

brittle transition
UF brittle-ductile transition
BT mechanical properties
RT ductility
phase transition

brittleness
BT mechanical properties
RT ductility

buckling

buildings
USE surface structures

bulk density
BT density
physical properties

bulk modulus
BT mechanical properties
RT compressibility
stress strain relations

bulking
RT dilatancy

buried pipes
BT pipes
RT pipe jacking
underground structures.....

buttress dams
BT dams
RT dam foundations

cable bolts
USE anchors

Cainozoic rocks
UF Cenozoic rocks
Quaternary rocks
Tertiary rocks
RT age
Mesozoic rocks
Palaeozoic rocks
Precambrian rocks

Cainozoic soils
UF Cenozoic soils
Quaternary soils
Tertiary soils
RT age
Mesozoic soils
Palaeozoic soils
Precambrian soils

caissons
USE cofferdams
OR piles (as appropriate)

calcareous rocks
USE carbonate rocks

calcareous soils
NT marls

calcite
BT minerals
RT carbonate rocks.....

calculation
USE computation

calibration

Cambrian
SEE Palaeozoic

cameras
RT photography.....

canals
USE hydraulic structures

capillarity
RT permeability
porosity
saturation

carbonaceous rocks
coaly rocks or rocks containing hydrocarbons
BT sedimentary rocks
NT anthracite
coal
lignite
oil shale
tar sands

carbonate rocks
UF calcareous rocks
NT chalk
dolomite
limestone
marble
RT calcite

Carboniferous
SEE Palaeozoic

case studies

casting

cataclasis
RT deformation

catastrophe theory
RT mathematical analysis.....

caverns
BT underground structures
NT storage caverns
RT cavities.....
excavations
openings

caving
BT mining
NT block caving

cavitation
RT water jets

cavities
for large man-made excavations use caverns
NT natural cavities
solution cavities
RT caverns.....

cavity location
BT location

cement
BT structural materials
RT concrete
mortar
plaster

cementation
RT grouting.....
gunite
shotcrete
support.....

cemented fill
BT fills

Cenozoic
SEE Cainozoic

centrifuges

ceramics

chalk
BT carbonate rocks
sedimentary rocks
RT dolomite
limestone

channels
BT hydraulic structures
surface structures

chemical analysis
RT geochemistry
thermal analysis

chemical changes
USE chemistry

chemical grouting
BT grouting

chemical properties
NT acidity
salinity

chemistry
UF chemical changes
RT acidity
adsorption
chemical analysis
chemical properties.....
corrosion
crystallization
diffusion
electroosmosis
evaporation
filtration
oxidation
polymers
salinity

chert
BT sedimentary rocks
NT taconite
RT quartz

chipping

chock supports
USE hydraulic props

circular
RT geometry
non circular

circular disks
USE disks

circular holes
BT holes

civil engineering

classification
NT rock classification
rock mass classification
soil classification

clastic rocks
BT sedimentary rocks
NT arenaceous rocks
argillaceous rocks
argillites
breccia
claystones
conglomerate
greywacke
gritstones
limestone
mudstone
oil shale
sandstones
siltstones
shales
tar sands
RT clays.....
phosphates

clay minerals
NT kaolin
montmorillonite

clays
UF quick clays
BT cohesive soils
NT bentonite
RT clastic rocks.....
expansive soils
feldspar
minerals.....
mud
sedimentary rocks.....
sediments
sensitive soils
shales
slates

claystones
BT clastic rocks
sedimentary rocks
RT argillaceous rocks

cleavage
tendency to split along planes
determined by the crystal structure
RT bedding
fissility
foliation
schistosity

cliffs
RT geomorphology

closure
USE convergence

cluster analysis
BT mathematical analysis
statistical analysis

coal
BT carbonaceous rocks
sedimentary rocks
NT anthracite
lignite
RT coal measures

coal measures
BT sedimentary rocks
RT coal.....
Palaeozoic rocks

coal mines
RT open pit mines
seams
underground mines

coal mining
BT mining
RT longwall mining
shortwall mining

coal spoil
BT waste materials

coasts
RT geomorphology

coatings

cofferdams
BT dams

cohesion
PREFER cohesionless soils
OR cohesive soils (if appropriate)
BT physical properties

cohesionless soils
RT gravel
sands.....

cohesive soils
NT bentonite
clays
RT loams
sensitive soils
silts

collapse
BT failure

collapsible soils
UF slump-prone soils
RT aeolian soils.....
residual soils.....
shrinkage
unsaturated soils

colloquia
USE conferences

columns
RT arches
pillars

comminution
RT crushing
grinding
mills.....
rock breaking
specific energy

compaction
PREFER soil compaction (if appropriate)

comparison

compressibility
BT mechanical properties
RT bulk modulus
compressive strength
consolidation tests
deformability
stress strain relations

compression
NT triaxial compression
uniaxial compression
RT pressure.....

compression strain

compression tests
BT field tests
lab tests
NT triaxial compression tests
uniaxial compression tests
RT Brazilian tests
indentation tests

compression waves
USE P waves

compressive strength
BT mechanical properties
RT compressibility
triaxial strength
uniaxial strength

compressive stress
BT normal stress

computation
UF calculation

computer programs
RT data processing

computers
RT data processing

concrete
BT structural materials
NT precast concrete
RT cement
gunite
mortar
shotcrete

conductivity
BT electrical properties
RT resistivity
superconductivity

cones
RT geometry

conferences

confining pressure
RT triaxial compression tests
triaxial tests.....

confining stress
USE confining pressure
(if appropriate)

conglomerate
BT clastic rocks
sedimentary rocks

congresses
USE conferences

consolidated drained triaxial tests
BT drained shear tests
drained triaxial tests
lab tests
shear tests
triaxial tests

consolidated undrained triaxial tests
BT lab tests
shear tests
triaxial tests
undrained shear tests
undrained triaxial tests

consolidation
RT overconsolidation
settlement
swelling

consolidation tests
BT lab tests
RT compressibility
oedometers
settlement

constitutive models (or laws)
USE stress strain relations

construction

contact area
RT friction.....

continuum mechanics
BT mechanics
RT discontinuum mechanics

contour blasting
USE controlled blasting

contracts
RT management
standards

control
UF remote control

controlled blasting
UF contour blasting
cushion blasting
line drilling
perimeter blasting
postsplitting
smooth blasting
BT blasting
RT presplitting

convergence
(of an underground opening) - the sides, or top and bottom, get closer together
UF closure
RT sag
squeezing ground

core logging

core recovery

core relief tests
USE doorstoppers
OR overcoring

cores
UF drill cores

correlation

corrosion
BT environmental effects
RT chemistry
weathering

costs
USE economics

Coulomb criterion
BT yield criteria

crack initiation
USE fracture initiation

crack patterns
USE fracture patterns

crack propagation
USE fracture propagation

cracking
PREFER fracture initiation
OR fracture propagation
(if appropriate)

cracks
RT fissures
fractures.....
microcracks
microfractures

craters
RT surface structures.....

creep
BT time dependent behaviour

creep tests
BT lab tests

Cretaceous
SEE Mesozoic

critical state
RT residual strength

cross-cuts
USE drifts

crosshole logging
BT borehole logging
NT crosshole seismic method

crosshole seismic method
BT borehole logging
crosshole logging
field tests
seismic methods
RT geophysics.....
seismic waves.....

crushability
BT mechanical properties

crushing
RT comminution
grinding
specific energy

crushing tests

cryogenic temperature
temperatures associated with liquid gases
RT freezing
low temperature
superconductivity

crystalline rocks
RT igneous rocks.....
metamorphic rocks.....

crystallization
RT chemistry
microstructure

crystallography
RT geology.....
microstructure

crystals
RT microstructure

cubes
RT geometry

culverts
USE hydraulic structures

cushion blasting
USE controlled blasting

cut and fill
BT mining
RT fills.....

cut offs
cut off walls or trenches
RT walls

cuts
RT excavations

cutting
USE rock cutting (if appropriate)

cuttings (excavations)
USE cuts

cyclic loading tests
BT lab tests
RT fatigue
time dependent behaviour.....

cylinders
RT geometry

dam foundations
BT foundations

damage

damping
RT absorption
attenuation
propagation
waves.....

dams
BT surface structures
NT arch dams
buttress dams
cofferdams
earth dams
embankment dams
gravity dams
rock fill dams
tailings dams
RT dam foundations
hydraulic structures.....
reservoirs

Darcys law

data acquisition

data logging
USE data acquisition

data processing
UF algorithms
RT computer programs
computers

debris
RT materials.....
rocks.....
soils.....
solifluction

deformability
UF deformation modulus
BT mechanical properties
RT stress strain relations

deformation
PREFER deformability (if appropriate)
UF distortion
RT cataclasis
strain.....

deformation cells
RT testing machines.....

deformation mechanisms

deformation modulus
USE deformability

degradation
BT environmental effects
RT denudation
erosion

densification
USE compaction
OR soil compaction

density
BT physical properties
NT bulk density

denudation
BT environmental effects
RT degradation
erosion

depth
UF shallowness

deserts
RT geomorphology

desiccation
RT dryness
environments

design

destressing
USE stress relief

determination
USE computation
OR measurement

detonation
RT blasting.....

detonation velocity
RT blasting.....

detonation waves
USE shock waves

development

Devonian
SEE Palaeozoic

dewatering
RT drainage

diabase
UF dolerite
BT basic rocks
igneous rocks
intrusive rocks

diamond

diamond drilling

diaphragm walls
BT surface structures
walls

dielectric constant
BT electrical properties

diffraction
RT propagation
waves.....

diffractometry

diffusion
UF transport
NT gas diffusion
moisture diffusion
RT chemistry
fluid flow.....

dilatancy
RT bulking

dilatation
USE dilatancy

dilatation waves
USE P waves

dilation
USE dilatancy

dilatometers
BT instruments

dimensional analysis
BT mathematical analysis
RT one dimensional analysis
two dimensional analysis
three dimensional analysis

diorite
BT igneous rocks
intermediate rocks
intrusive rocks

dip
RT inclination
strike

dip direction

dip slip faults
BT faults

direct shear tests
BT lab tests
shear tests
NT drained direct shear tests
undrained direct shear tests
RT ring shear tests

discontinuities
UF notches
RT faults.....
fissures
joints

discontinuum mechanics
UF mechanics of discontinua
BT mechanics
RT continuum mechanics

disks
UF circular disks

dispersion
RT propagation
waves.....

displacement
RT distance
hole spacing
joint spacing

displacement discontinuity method
USE boundary integral method

distance
RT displacement
hole spacing
joint spacing

distance measurement
BT field tests
RT surveying.....

distortion
USE deformation

distortion waves
USE S waves

distribution laws
RT mathematical analysis.....
probability
statistical analysis.....

dolerite
USE diabase

dolomite
BT carbonate rocks
sedimentary rocks
RT chalk
limestone

doorstoppers
RT overcoring
strain relief
stress relief

dowels (reinforcing members)
USE bolts

downhole drills

drainage
RT dewatering
hydrology
leakage
percolation
seepage

drained direct shear tests
BT direct shear tests
drained shear tests
lab tests
shear tests

drained shear tests
BT lab tests
shear tests
NT consolidated drained triaxial tests
drained direct shear tests
drained triaxial tests

drained triaxial tests
BT drained shear tests
lab tests
shear tests
triaxial tests
NT consolidated drained triaxial tests

drawdown
RT hydrology

dredging

drifts
UF adits
crosscuts
galleries (mines)
haulages
levels
roadways (underground)
BT underground structures
RT driving
rock passes
support.....
underground mines

drill bits

drill cores
USE cores

drill holes
BT holes

drill rigs

drillability
BT mechanical properties
RT borability
penetration rate

drilling
NT air drilling
auger drilling
diamond drilling
exploration drilling
hydraulic drilling
jet drilling
percussion drilling
rotary drilling

drilling (contd)
thermal drilling
turbodrilling

drilling fluids

drilling muds
USE drilling fluids

drilling rate

drills
NT downhole drills

drills and drilling
USE drilling

driving
constructing a drift
RT drifts
mining.....
tunnelling.....

dryness
RT desiccation
environments

ductility
BT mechanical properties
RT brittle materials
brittle transition
brittleness

dumps
UF spoil heaps
waste heaps/dumps
BT surface structures
RT berms
embankments
slopes.....

dunite
BT basic rocks
igneous rocks
intrusive rocks

durability
USE slake durability tests
(if appropriate)

dust
RT environments

dykes
tabular, igneous intrusions
(not embankments or levees)
RT igneous rocks.....
intrusive rocks.....

dynamic analysis
BT mathematical analysis

dynamic compaction/consolidation
USE soil compaction
AND dynamics

dynamic modulus
USE dynamic properties

dynamic properties
BT mechanical properties
RT seismic waves.....
shock effects

dynamic tests
BT field tests
lab tests
NT resonant column tests

dynamics
BT mechanics
NT hydrodynamics

dynamometers
BT instruments

earth crust
RT plate tectonics
tectonics

earth dams
BT dams
embankment dams
RT dam foundations
gravity dams

earth fill
BT fills
RT reinforced earth

earth masses
USE soil masses

earth movement
USE ground movement

earth pressure

earthquake mechanisms

earthquake prediction

earthquake triggering

earthquakes
RT accelerograms
aftershocks
foreshocks
ground movement
microseisms
seismicity
shock effects
strata movement
tsunami

economics
UF costs

education

effective stress
RT pore pressure
total stress

elastic analysis
BT mathematical analysis

elastic constant
USE elastic properties

elastic models
RT model tests
photoelastic models

elastic modulus
USE Youngs modulus

elastic parameters
USE elastic properties

elastic properties
UF elastic constant
elastic parameters
BT mechanical properties
RT stress strain relations
viscoelasticity

elastic waves
PREFER a narrower term (if appropriate)
NT body waves
Love waves
P waves
Rayleigh waves
S waves
seismic waves
surface waves
RT acoustic waves
inelastic waves.....
ultrasonic waves

elasticity
PREFER elastic properties (if appropriate)
BT mechanics
RT inelasticity
photoelasticity
plasticity
Poissons ratio
viscoelasticity
Youngs modulus

elastomers
RT plastics
rubbers

electrical energy

electrical methods
includes, but is not limited to, geoelectrical methods of exploration
NT electromagnetic methods
electromagnetic sounding
induced polarisation
magnetotelluric methods
resistivity methods
telluric methods
RT analogues
geoelectricity
geophysics.....
sounding

electrical properties
BT physical properties
NT conductivity
dielectric constant
resistivity

electricity
NT geoelectricity
piezoelectricity
RT electronics
superconductivity

electromagnetic methods
BT electrical methods
NT electromagnetic sounding
induced polarisation
RT geophysics.....
mapping
radar

electromagnetic sounding
BT electrical methods
electromagnetic methods
field tests
sounding
RT geophysics.....

electromagnetic sounding (contd)
mapping
radar

electromagnetic wave emission

electromagnetic waves
NT gamma rays
infrared
microwaves
x rays

electromagnetism
BT magnetism
RT superconductivity

electron microscopy
BT microscopy

electronics
RT electricity

electroosmosis
RT chemistry
fluid flow.....

ellipses
RT geometry

ellipsoids
RT geometry

elongation

embankment dams
BT dams
NT earth dams
rock fill dams
tailings dams
RT dam foundations
gravity dams

embankment foundations
BT foundations

embankments
BT surface structures
RT berms
dumps
slopes.....

energy
NT electrical energy
geothermal energy
specific energy
strain energy
surface energy
thermal energy

energy methods
BT mathematical analysis

energy release rate
RT surface energy

engineering geology
BT geology

engineering geology maps
BT maps

environmental effects
effects of the environment
NT corrosion
degradation
denudation
erosion
oxidation
soil formation
weathering
RT pollution

environments
RT air
compression.....
dryness
dust
pollution
pressure.....
temperature.....
tension

equation of state
RT mathematical analysis.....

equipment
RT machines.....

erosion
BT environmental effects
RT degradation
denudation

error analysis
BT mathematical analysis

evaporation
RT chemistry

evaporitic rocks
BT sedimentary rocks
NT anhydrite
gypsum
potash
rock salt

excavations
PREFER a more specific term (if appropriate)
RT caverns.....
foundations.....
openings
surface structures.....
underground structures.....

exfoliation

expansion
RT shrinkage
volume
volume strain

expansion tests
BT lab tests

expansive soils
RT clays
shrinkage
swelling
water content

experimental techniques
USE field tests
OR lab tests

exploration

exploration drilling

explosions
RT shock effects

explosive charges

explosives
NT slurry explosives
RT blasting.....

extension tests
BT lab tests
RT tensile strength

extensometers
BT instruments
RT monitoring systems

extraction
RT mining.....
pillar extraction

extrusive rocks
UF volcanic rocks
BT igneous rocks

extrusive rocks (contd)
NT andesite
basalt
ignimbrite
lava
rhyolite
tuff

fabrics
USE microstructure
OR textiles

faces
UF working faces
RT mines.....
open pit mines
underground mines
underground structures.....

factor analysis
BT mathematical analysis
statistical analysis

failure
NT brittle failure
collapse
plane shear failure
rotational failure
shear failure
toppling failure
wedge failure
RT post failure behaviour

failure analysis
BT mathematical analysis

failure criteria
UF failure surface (mathematical)
NT Coulomb criterion
Griffith criterion
yield criteria
RT brittle failure

failure initiation
RT fracture initiation

failure mechanisms

failure mode
USE failure mechanisms

failure prediction

failure surface
the physical surface of failure

failure surface (mathematical)
USE failure criteria

fatigue
BT mechanical properties
time dependent behaviour
RT cyclic loading tests
fracture toughness

fault gouge

fault mechanisms

faults
NT dip slip faults
normal faults
oblique slip faults
overthrust faults
strike slip faults
thrust faults
RT discontinuities
fault mechanisms
fissures
joints

feasibility

feldspar
BT minerals
RT anorthosite
plagioclase
silicates

field tests
NT borehole logging
borehole shear tests
crosshole logging
crosshole seismic method
distance measurement
dynamic tests
electromagnetic sounding
flat jack tests
heater tests
index tests
jacking tests
loading tests
plate bearing tests
pressure tests
pullout tests
pumping tests
screw plate tests
seismic methods
shear tests
sounding
vane shear tests
RT surveying.....

filling
UF stowing

fills
BT structural materials
NT backfill
cemented fill
earth fill
hydraulic fill
rock fill
sand fill
RT fly ash
packing.....
support.....
waste materials.....

filtration
RT chemistry

finite difference method
BT mathematical analysis
numerical analysis

finite element analysis
BT mathematical analysis
numerical analysis

finite strain
RT deformation

fissility
tendency to split along relatively
smooth planes parallel to the bedding
RT bedding
cleavage
foliation
schistosity

fissures
RT discontinuities
faults
joints

flat jack tests
BT field tests
jacking tests
RT stress measurement

flat jacks
BT jacks

flexural slip
RT slip

floors
RT underground structures.....

flow pattern
RT hydraulics.....

flow rate
RT hydraulics.....

fluid flow
UF gas flow
hydraulic flow
NT groundwater flow
water flow
RT diffusion
electroosmosis
hydraulics.....

fluid mechanics
USE hydraulics

fluid pressure
USE hydraulic pressure

fluid properties
USE hydraulic properties

fluids
RT hydraulics.....

fly ash
BT structural materials
waste materials
RT fills.....

flysch
UF molasse
BT sedimentary rocks
RT shales

fold mechanisms

folds
RT kink band

foliation
RT cleavage
fissility
layered systems
layers
schistosity
strata

footings
BT foundations

footwalls
RT hanging walls
mines.....
open pit mines
underground mines
walls

foreshocks
RT aftershocks
earthquakes

fossils
RT palaeontology

foundations
NT bridge foundations
dam foundations
embankment foundations
pile foundations
raft foundations
RT abutments
anchorages
base courses
footings
mats
piles
rafts
springs
surface structures.....
underpinning

Fourier analysis
BT mathematical analysis
statistical analysis

fracture direction
USE fracture orientation

fracture initiation
UF crack initiation

fracture mechanics
USE fracture mechanisms

fracture mechanisms

fracture orientation
UF fracture direction

fracture patterns
UF crack patterns

fracture propagation
UF crack propagation

fracture toughness
BT mechanical properties
RT fatigue

fracture zones

fractures
NT shear fractures
tensile fractures
RT brittle fracture
cracks

fractures (contd)
RT fissures
microcracks
microfractures

fracturing
USE rock breaking (if appropriate)

fragmentation
RT blasting.....
rock breaking

freezing
RT cryogenic temperature
frost
frozen ground
frozen rocks
frozen soils
ice
low temperature

friction
NT internal friction
pile friction
residual friction
RT contact area
friction angle
stick slip
surface properties.....

friction angle
RT internal friction

friction coefficient
USE friction angle

friction tests
BT lab tests

fringes
USE Moire fringes

frost
RT freezing
low temperature

frost heave
BT heave

frozen ground
PREFER frozen rocks
OR frozen soils (if appropriate)
RT freezing
ice
low temperature
permafrost

frozen rocks
RT freezing
low temperature
permafrost

frozen soils
NT permafrost
RT freezing
low temperature

gabbro
BT basic rocks
igneous rocks
intrusive rocks
NT norite
RT basalt

gabions
RT hydraulic structures.....
rock fill

galleries (mines)
USE drifts

gamma rays
BT electromagnetic waves

gas absorption
BT absorption

gas burst
RT rock burst

gas diffusion
BT diffusion

gas flow
USE fluid flow

gases

gauges
USE instruments

geochemistry
RT chemical analysis

geodesy
RT geophysics.....

geoelectricity
BT electricity
geophysics

geology
NT engineering geology
glaciology
sedimentology
structural geology
RT crystallography

geomagnetism
BT geophysics
RT piezomagnetism

geomechanics
use only for very general material, prefer a narrower term if possible
BT geophysics
mechanics
NT rock mechanics
soil mechanics

geometry
RT circular
cones
cubes
cylinders
ellipses
ellipsoids
helical
non circular
rectangular
shape
spheres
spiral

geomorphology
RT cliffs
coasts
deserts
karst
mountains
rivers
sea bed
topography

geophones
BT instruments
RT monitoring systems

geophysics
NT geoelectricity
geomagnetism
geomechanics
rock mechanics
seismology
soil mechanics
RT electrical methods.....
geodesy
gravimetry

geophysics (contd)
RT induced polarisation
seismic methods.....
sounding.....
tectonics.....

geotechnical classification
USE rock mass classification

geotechnical properties
USE mechanical properties

geothermal energy
BT thermal energy

glacial till
UF moraine (material)
till
BT soils

glaciers

glaciology
BT geology

glass

gneiss
BT metamorphic rocks

gold mines
RT underground mines

graben
BT rock masses

grain boundaries
BT boundaries

grain orientation

grain size
RT size distribution
size effect

grains
RT interlocking
microstructure

granite
BT acidic rocks
igneous rocks
intrusive rocks

granodiorite
BT igneous rocks
intermediate rocks
intrusive rocks

granular materials
RT aggregates
gravel
sands

graph data
RT graphical analysis

graphical analysis
BT mathematical analysis
RT accelerograms
graph data
hodographs
holography
isobars
isochromatics
isoclinics
isopachs
isostatics
mapping
maps
nomograms
slip lines
stereographic projection

gravel
RT aggregates
cohesionless soils
sands
sediments

gravimetry
RT geophysics.....

gravity

gravity dams
BT dams
RT dam foundations
earth dams
embankment dams
rock fill dams
tailings dams

greywacke
BT clastic rocks
sandstones
sedimentary rocks

Griffith crack theory

Griffith criterion
BT failure criteria
RT brittle failure

grindability
BT mechanical properties

grinding
UF milling
RT comminution
crushing
mills.....
specific energy

gritstones
BT clastic rocks
sedimentary rocks
RT sandstones

ground movement
UF earth movement
rock movement
soil movement
RT earthquakes
rock burst
settlement
strata movement
subsidence

ground stress

groundwater
RT aquifers
hydrology
water table
wells

groundwater flow
BT fluid flow
water flow

grouted bolts
BT bolts
support
NT resin grouted bolts
RT resins

grouted rebars, dowels etc
USE grouted bolts

grouting
BT support
NT chemical grouting
RT cementation
injection

gunite
RT cementation
concrete.....
shotcrete
support.....

gypsum
BT evaporitic rocks
sedimentary rocks
RT anhydrite

half spaces
RT mathematical analysis.....

halite
USE rock salt

handbooks

hanging walls
RT footwalls
mines.....
open pit mines
underground mines
walls

harbours
BT hydraulic structures
surface structures
RT piers

hard rock tunnelling
BT tunnelling
RT New Austrian Tunnelling Method
soft ground tunnelling
tunnelling machines

hard rocks

hardness
BT mechanical properties
NT indentation hardness
RT abrasiveness
surface properties.....

haulages
USE drifts

hazards
RT accidents
safety

heat
RT temperature.....
thermodynamics

heat flow
USE heat transfer

heat transfer
RT thermal conductivity

heater tests
BT field tests

heave
NT frost heave
RT time dependent behaviour.....
uplift

height

helical
RT geometry
spiral

heterogeneity
USE inhomogeneity

high pressure

high temperature

highways
USE roads

hodographs
RT graphical analysis

hole spacing
RT displacement
joint spacing

holes
NT boreholes
circular holes
drill holes
RT slots

hollow cylinder tests
BT lab tests
RT biaxial tests
tensile strength
torsion tests

holography
RT graphical analysis

homogeneity
RT inhomogeneity

horizontal

humidity
RT environments
moisture
water

hydraulic cells
RT testing machines.....

hydraulic conductivity
RT permeability

hydraulic drilling

hydraulic fill
BT fills

hydraulic flow
  USE fluid flow

hydraulic fracturing
  UF hydrofracturing

hydraulic pressure
  UF fluid pressure
  NT hydrostatic pressure
    joint water pressure
    water pressure
  RT pore pressure

hydraulic properties
  UF fluid properties
  BT physical properties
  NT thixotropy
    turbidity
    viscosity
  RT permeability
    porosity
    storage

hydraulic props
  UF chock supports
    powered supports
    shield supports
  BT props
    support

hydraulic structures
  UF canals
    culverts
  NT channels
    harbours
    locks
    reservoirs
  RT dams.....
    gabions
    surface structures.....

hydraulic systems
  USE hydraulics

hydraulics
  UF fluid mechanics
  NT hydrodynamics
    hydrology
    hydrostatics
  RT flow pattern
    flow rate
    fluid flow.....
    fluids

hydrodynamics
  BT dynamics
    mechanics

hydrofracturing
  USE hydraulic fracturing

hydrogeology
  USE hydrology

hydrography
  USE hydrology

hydrology
  RT aquifers
    drainage
    drawdown
    leakage
    percolation
    seepage
    wells

hydrostatic loading
  BT loading.....

hydrostatic pressure
  BT hydraulic pressure

hydrostatic pressure chambers
  BT testing machines
  RT pressure cells
    pressure tests

hydrostatic stress
  a state of stress where the three principal stresses are equal

hydrostatics

hysteresis
  RT mechanical properties.....
    stress strain relations

ice
  RT frozen ground
    permafrost
    snow

igneous rocks
  NT acidic rocks
    andesite
    basalt
    basic rocks
    diabase
    diorite
    dunite
    extrusive rocks
    gabbro
    granite
    granodiorite

igneous rocks (contd)
NT ignimbrite
intermediate rocks
intrusive rocks
lava
norite
porphyry
quartz monzonite
rhyolite
serpentinite
tonalite
tuff
RT breccia
crystalline rocks
dykes
magma
volcanoes

ignimbrite
BT extrusive rocks
igneous rocks
RT volcanoes

images
USE airborne photography
OR photography
OR remote sensing

impact

impact tests
BT lab tests

impregnation

impression packers

impulses
USE pulses

in situ stress

in situ stress measurement
USE in situ stress
AND stress measurement

inclination
UF steepness
RT dip

inclined shafts
BT shafts
underground structures
RT drifts

inclinometers
BT instruments
RT boreholes
monitoring systems
tiltmeters

inclusions
NT rigid inclusions
RT microstructure

indentation hardness
BT hardness

indentation tests
BT lab tests
RT anisotropy
compression tests.....
microstructure
punch tests

index tests
BT field tests
lab tests

induced polarisation
BT electrical methods
electromagnetic methods
RT exploration
geophysics.....

inelastic waves
NT plastic waves
shock waves
RT elastic waves.....
seismic waves.....

inelasticity
BT mechanical properties
RT elasticity
plasticity

information retrieval

infra red
BT electromagnetic waves

inhomogeneity
UF heterogeneity
RT homogeneity

injection
RT grouting.....

instruments
PREFER a narrower term (if appropriate)
UF gauges
NT accelerometers
borehole gauges
dilatometers
dynamometers
extensometers
geophones
inclinometers
load cells
penetrometers
permeameters
photoelastic gauges
piezometers
pressuremeters
seismometers
strain gauges
stress meters
tiltmeters
transducers
vibrating wire instruments
RT monitoring systems

integral sampling
BT field tests
RT sampling

interfaces
RT boundaries.....
mathematical analysis.....

interferometry
RT lab tests.....
Moire fringes

interlocking
RT blocks
grains
jointed rocks

intermediate rocks
on the basis of silica content, intermediate between acidic rocks and basic rocks
BT igneous rocks
NT andesite
diorite
granodiorite
quartz monzonite
tonalite

internal friction
BT friction
RT friction angle
shear strength

interstitial fluids
USE pore fluids

intrusive rocks
UF plutonic rocks
BT igneous rocks
NT diabase
diorite
dunite
gabbro
granite
granodiorite
norite
quartz monzonite
tonalite
RT dykes
extrusive rocks.....
magma

isobars
RT graphical analysis
principal stress
stress field

isochromatics
RT graphical analysis
photoelastic tests
shear stress
stress field

isoclinics
RT graphical analysis
photoelastic tests
principal stress
stress field

isopachs
RT analogues
graphical analysis
stress field

isostatics
RT graphical analysis
principal stress
stress field

isotropy
RT anisotropy

jacking tests
NT flat jack tests

jacks
NT flat jacks
RT pipe jacking

jet drilling

joint continuity
UF joint persistence

joint filling

joint formation

joint orientation

joint patterns

joint persistence
USE joint continuity

joint roughness
USE joint surface

joint sets
UF sheeting

joint shear strength

joint spacing

joint surface
UF joint roughness

joint water pressure
BT hydraulic pressure
water pressure

jointed rocks
BT rock masses
RT massive rocks
interlocking

joints
RT discontinuities
fissures

Jurassic
SEE Mesozoic

kaolin
BT clay minerals

karst

kinematics
BT mechanics

kinetics
BT mechanics

kink band
RT folds

lab tests
NT bending tests
biaxial tests
Brazilian tests
compression tests
consolidated drained triaxial tests
consolidated undrained triaxial tests
creep tests
cyclic loading tests
direct shear tests
drained direct shear tests
drained shear tests
drained triaxial tests
expansion tests
extension tests
friction tests
hollow cylinder tests
impact tests
indentation tests
model tests
photoelastic tests
point load tests
punch tests
ring shear tests
shear tests
simple shear tests
slake durability tests
slaking tests
tension tests
torsion tests
triaxial compression tests
triaxial tests
unconsolidated undrained triaxial tests
undrained direct shear tests
undrained shear tests
undrained triaxial tests
uniaxial compression tests
uniaxial tension tests
uniaxial tests
vane shear tests
x ray analysis
RT interferometry
microscopy.....
Moire fringes
thin sections

lamellae
USE laminae

laminae
  RT foliation
    layers

landslides
  RT avalanches
    rock slides
    slide mechanisms
    slip

Laplace transformation
  RT mathematical analysis

lasers

laterites
  BT residual soils

lava
  BT extrusive rocks
    igneous rocks
  RT volcanoes

laws
  UF legislation

layered systems
  RT foliation
    strata

layers
  PREFER layered systems (if appropriate)
  RT foliation
    laminae
    strata

leaching

leakage
  RT drainage
    hydrology
    percolation
    seepage
    water flow.....

least squares method
  BT mathematical analysis

legislation
  USE laws

length

lenticular orebodies
  BT orebodies

levelling
  USE surveying

levels (in a mine)
  USE drifts

lignite
  BT carbonaceous rocks
    coal
    sedimentary rocks
  RT anthracite

lime
  BT minerals

limestone
  BT carbonate rocks
    sedimentary rocks
  RT aggregates
    chalk
    dolomite
    karst
    marls
    marlstone

limit analysis
  BT mathematical analysis

limit equilibrium method
  BT mathematical analysis

line drilling
  USE controlled blasting

linearity
  RT nonlinearity

linings
  NT borehole linings
    shaft linings
    tunnel linings
  RT support.....

liquefaction

liquid limit
  RT plasticity

liquids
  RT fluids

literature surveys
  USE reviews

lithology
  RT microstructure
    petrography
    petrology

load cells
BT instruments
RT monitoring systems

loading
NT hydrostatic loading
preloading
RT unloading

loading rate
RT time dependent behaviour.....

loading tests
may be applied to eg, tests of foundations, piles etc.
PREFER a more specific term (listed under one of the BTs)
BT field tests
lab tests

loams
BT organic soils
RT cohesive soils.....

location
UF position
NT cavity location
shaft location

locks
BT hydraulic structures
surface structures

loess
BT aeolian soils
RT sediments
silts

logging
USE borehole logging
OR core logging

longitudinal waves
USE P waves (if appropriate)

longwall mining
BT mining
RT coal mining

loosened zones
RT blasting.....

Love waves
BT elastic waves
seismic waves
surface waves
RT Rayleigh waves

low temperature
RT cryogenic temperature
freezing
frost
frozen ground

Lugeon tests
RT permeability tests

lunar rocks

lunar soils

machines
NT bolting machines
boring machines
hydrostatic pressure chambers
mining machines
stiff testing machines
testing machines
tunnelling machines
RT equipment

magma
RT igneous rocks.....
intrusive rocks.....
molten rocks

magnetic properties
BT physical properties

magnetism
NT electromagnetism
RT geomagnetism

magnetite
BT minerals
ores
RT aggregates

magnetotelluric methods
BT electrical methods
RT telluric methods

management
RT contracts
standards

mapping
RT electromagnetic methods.....
resistivity methods
surveying.....

maps
NT engineering geology maps

marble
BT carbonate rocks
metamorphic rocks

marls
BT calcareous soils
carbonate rocks
sedimentary rocks
rocks
soils
RT argillaceous rocks
limestone

marlstone
BT carbonate rocks
clastic rocks
sedimentary rocks
RT limestone

masonry
RT structural materials.....

massive orebodies
BT orebodies

massive rocks
non-discontinuous rocks
BT rock masses
RT jointed rocks

materials
use in a general sense only; prefer
a more specific term (if appropriate)
NT aggregates
alloys
backfill
bitumen
brittle materials
cement
cemented fill
ceramics
coal spoil
concrete
debris
earth fill
elastomers
fills
fly ash
glass
granular materials
hydraulic fill
metals
model materials
mortar
overburden
photoelastic materials
plaster
plastics

materials (contd)
NT porous materials
precast concrete
reinforced earth
resins
rock fill
rock like materials
salts
sand fill
steel
structural materials
textiles
waste materials
wire mesh
wood
RT additives
coatings
mixtures
pastes
powders
slurries

mathematical analysis
UF theoretical studies
theories
NT back analysis
boundary integral method
cluster analysis
dimensional analysis
dynamic analysis
elastic analysis
energy methods
error analysis
factor analysis
failure analysis
finite difference method
finite element analysis
Fourier analysis
graphical analysis
least squares method
limit analysis
limit equilibrium method
mathematical models
Monte Carlo method
multivariate analysis
numerical analysis
one dimensional analysis
regression analysis
simulation
slice analysis
slope analysis
stability analysis
statistical analysis
strain analysis
stress analysis
structural analysis
three dimensional analysis
two dimensional analysis

mathematical analysis (contd)
RT boundaries
catastrophe theory
distribution laws
equation of state
half spaces
hodographs
interfaces
isobars
isochromatics
isoclinics
isopachs
isostatics
Laplace transformation
matrices
nomograms
perturbation theory
probability
slip lines
stereographic projection
symmetry
tensile cut off
wave equation

mathematical models
BT mathematical analysis
RT simulation

matrices
RT mathematical analysis.....

mats
RT foundations

measurement

measuring bolts
BT bolts

mechanical properties
UF geotechnical properties
BT physical properties
NT blastibility
borability
brittle transition
brittleness
bulk modulus
compressibility
compressive strength
crushability
deformability
drillability
ductility
dynamic properties
elastic properties
fatigue
fracture toughness

mechanical properties (contd)
grindability
hardness
plasticity
Poissons ratio
residual strength
rigidity
rippability
rock quality designation
shear modulus
shear strength
strength
tensile strength
triaxial strength
uniaxial strength
yield criteria
Youngs modulus
RT hysteresis
strength reduction

mechanics
NT continuum mechanics
discontinuum mechanics
dynamics
elasticity
geomechanics
hydrodynamics
hydrostatics
kinematics
kinetics
rheology
rock mechanics
soil mechanics
statics

mechanics of discontinua
USE discontinuum mechanics

mechanisation
USE automation

mechanism
USE mechanics
OR deformation mechanisms
earthquake mechanisms
failure mechanisms
fault mechanisms
fold mechanisms
fracture mechanisms
slide mechanisms

melting

membranes
RT triaxial compression tests
triaxial tests.....

Mesozoic rocks
UF Cretaceous rocks
Jurassic rocks
Triassic rocks
RT age
Cainozoic rocks
Palaeozoic rocks
Precambrian rocks

Mesozoic soils
UF Cretaceous soils
Jurassic soils
Triassic soils
RT age
Cainozoic soils
Palaeozoic soils
Precambrian soils

metals
any metallic element may be used as a keyword
NT alloys
steel

metamorphic rocks
UF altered rocks
NT argillites
gneiss
marble
quartzites
schist
serpentinite
slates
RT breccia
crystalline rocks
mica

metamorphism
USE metamorphic rocks
OR shape

methods

metric units

metros
USE railways

mica
BT minerals
RT muscovite
silicates

microcracks
RT cracks

microfractures
RT fractures

microscopy
NT electron microscopy
RT lab tests.....

microseismic methods
USE acoustic emission

microseisms
RT earthquakes

microstructure
UF fabrics
petrofabrics
RT crystallization
crystallography
crystals
grains
inclusions.....
indentation tests
lithology
rigid inclusions
structural geology
texture

microwaves
BT electromagnetic waves

milling
USE grinding

mills
NT ball mills
rod mills
RT comminution
grinding

mineralogy

minerals
NT anorthosite
bauxite
calcite
clay minerals
diamond
feldspar
kaolin
lime
magnetite
mica
montmorillonite
muscovite
olivine
ores

minerals (contd)
plagioclase
pyrite
quartz
sylvinite
RT silicates
sulphides

mines
NT coal mines
gold mines
open pit mines
salt mines
underground mines
RT drifts
faces
footwalls
hanging walls
openings
quarries
seams
shafts.....
support.....
tunnels.....
underground structures.....

mining
NT bord and pillar
block caving
caving
coal mining
cut and fill
longwall mining
open stoping
room and pillar
shortwall mining
stoping
trackless mining
undersea mining
RT boring
driving
extraction
pillar extraction
raising
reaming
scaling
shaft sinking
undercutting
underpinning

mining machines
BT machines

Mississippian
SEE Palaeozoic

mixing

mixtures

model materials
RT rock like materials

model studies
USE model tests
OR models

model tests
PREFER mathematical models (if appropriate)
BT lab tests
RT elastic models
photoelastic models
resin models
scale models

models
PREFER mathematical models (if appropriate)
NT elastic models
photoelastic models
resin models
rheologic models
scale models

modulus of rigidity
USE shear modulus

Mohr theory

Moire fringes
UF fringes
RT interferometry
lab tests.....
photoelastic tests

moisture
RT environments
humidity
water

moisture absorption
BT absorption

moisture diffusion
BT diffusion

molasse
USE flysch

molten rocks
RT magma

monitoring systems
RT accelerometers
borehole gauges
extensometers
geophones
inclinometers
instruments.....
load cells
permeameters
piezometers
seismometers
strain gauges
stress meters
tiltmeters
transducers
vibrating wire instruments

monitors (water jets)
USE water jets

Monte Carlo method
BT mathematical analysis
statistical analysis

montmorillonite
BT clay minerals

moraine (material)
USE glacial till

morphology
USE geomorphology
OR shape
(as appropriate)

mortar
BT structural materials
RT cement
concrete.....
plaster

motion

mountains
UF orogeny
RT geomorphology

mud
RT clays.....
drilling fluids
grouting.....

mudflows
USE solifluction

mudrocks
USE sedimentary rocks

mudstone
BT clastic rocks
sedimentary rocks
RT argillaceous rocks

multivariate analysis
BT mathematical analysis
statistical analysis

muscovite
BT minerals
RT mica
silicates

muskeg
RT soils

natural cavities
BT cavities
RT caverns.....
solution cavities

natural slopes
BT slopes
RT rock slopes

New Austrian Tunnelling Method
BT tunnelling
RT hard rock tunnelling
soft ground tunnelling

noise
NT rock noise
RT acoustic emission
acoustics
vibration

nomenclature

nomograms
RT graphical analysis
mathematical analysis

non circular
RT circular
geometry

nondestructive tests
NT radiographic tests
ultrasonic tests

nonlinearity
RT linearity

norite
BT basic rocks
gabbro
igneous rocks
intrusive rocks

normal faults
BT faults

normal stress
NT compressive stress
tensile stress
RT shear stress

notches
USE discontinuities

nuclear waste
USE radioactive waste

numerical analysis
BT mathematical analysis
NT boundary integral method
finite difference method
finite element analysis

oblique slip faults
BT faults

oceans
RT tides

oedometers
RT consolidation tests
settlement

offshore structures
BT structures
RT surface structures.....

oil shale
BT carbonaceous rocks
clastic rocks
sedimentary rocks
shales
RT tar sands

oil technology
USE petroleum technology

olivine
BT minerals
RT serpentinite

one dimensional analysis
BT mathematical analysis
RT dimensional analysis
two dimensional analysis
three dimensional analysis

open cast mines
USE open pit mines

open pit mines
RT coal mines
quarries
underground mines

open stoping
BT mining
stoping

openings
PREFER underground structures (if general) or use a more specific term
RT caverns.....
excavations
mines.....
underground structures.....

optical methods
RT diffractometry
interferometry
lasers
microscopy.....
Moire fringes
petroscopes
photography
polariscopes

optical properties
BT physical properties

Ordovician
SEE Palaeozoic

ore passes
USE rock passes

orebodies
BT rock masses
NT lenticular orebodies
massive orebodies
tabular orebodies
RT ores
seams

ores
BT minerals
NT bauxite
magnetite
sylvinite
RT orebodies.....

organic soils
NT loams
peat

orientation
RT dip direction
strike

orogeny
USE mountains

overburden
BT waste materials
RT structural materials.....

overconsolidation
RT consolidation

overcoring
RT doorstoppers
strain relief
stress relief

overthrust faults
BT faults

oxidation
BT environmental effects
RT chemistry

P waves
UF compression waves
dilatation waves
longitudinal waves
pressure waves
primary waves
BT body waves
elastic waves
seismic waves
RT S waves

packing
BT support
NT anhydrite packing
RT fills.....
props.....

palaeontology
RT fossils

Palaeozoic rocks
UF Cambrian rocks
Carboniferous rocks
Devonian rocks
Mississippian rocks
Ordovician rocks
Pennsylvanian rocks
Permian rocks
Silurian rocks
RT age
Cainozoic rocks
Mesozoic rocks
Precambrian rocks

Palaeozoic soils
UF Cambrian soils
Carboniferous soils
Devonian soils
Mississippian soils
Ordovician soils
Pennsylvanian soils
Permian soils
Silurian soils
RT age
Cainozoic soils
Mesozoic soils
Precambrian soils

pastes
RT materials.....

patents

patterns
RT fracture patterns
joint patterns

pavements
UF aircraft pavements
BT surface structures
RT base courses
foundations.....
roads
subgrades

peat
BT organic soils

pebbles
BT rocks

penetration

penetration rate
RT borability
drillability
performance
tunnelling.....
tunnelling machines.....

penetration tests

penetrometers
BT instruments

Pennsylvanian (age)
SEE Palaeozoic

percolation
RT drainage
hydrology
leakage
seepage

percussion drilling

performance
RT accuracy
failure
penetration rate
reliability
response
sensitivity
stability.....

perimeter blasting
USE controlled blasting

permafrost
BT frozen soils
RT frozen ground
frozen rocks
ice

permeability
UF relative permeability
RT capillarity
hydraulic conductivity
porosity
pumping tests
saturation

permeability tests
NT pumping tests
RT Lugeon tests

permeameters
BT instruments
RT monitoring systems

Permian
SEE Palaeozoic

perturbation theory
RT mathematical analysis.....

petrofabrics
USE microstructure

petrography
RT lithology

petroleum technology
UF oil technology

petrology
RT lithology

petroscopes
RT boreholes
optical methods

phase transition
RT brittle transition

photo interpretation
RT airborne photography

photoelastic gauges
BT instruments

photoelastic materials

photoelastic models
RT elastic models
model tests

photoelastic tests
BT lab tests
RT isochromatics
isoclinics
Moire fringes

photoelasticity
RT elasticity

photogrammetry
RT field tests.....
lab tests.....
photography.....

photography
NT airborne photography
RT cameras
photogrammetry

physical properties
NT blastability
borability
brittleness
brittle transition
bulk density
bulk modulus
cohesion
compressibility
compressive strength
conductivity

physical properties (contd)
NT crushability
deformability
density
dielectric constant
drillability
ductility
dynamic properties
elastic properties
electrical properties
fatigue
fracture toughness
grindability
hardness
hydraulic properties
magnetic properties
mechanical properties
optical properties
plasticity
Poissons ratio
residual strength
resistivity
rigidity
rippability
rock quality designation
shear modulus
shear strength
strength
tensile strength
thermal conductivity
thermal diffusivity
thermal properties
thixotropy
triaxial strength
turbidity
uniaxial strength
viscosity
yield criteria
Youngs modulus

physics
NT acoustics
continuum mechanics
discontinuum mechanics
dynamics
elasticity
electricity
electromagnetism
electronics
geoelectricity
geomagnetism
geomechanics
geophysics
heat
hydraulics
hydrodynamics
hydrostatics

physics (contd)
NT kinematics
kinetics
magnetism
mechanics
piezoelectricity
radioactivity
rheology
rock mechanics
soil mechanics
statics
thermodynamics
ultrasonics

picks
RT rock cutting

piers
PREFER piles (if appropriate)
UF wharves
BT surface structures
RT anchorages
bridges
harbours

piezoelectricity
BT electricity
physics

piezomagnetism
RT geomagnetism

piezometers
BT instruments
RT monitoring systems

pile driving

pile foundations
BT foundations

pile friction
BT friction
soil structure interaction

piles
RT foundations.....
sheet piles

piling
USE pile driving
OR piles

pillar extraction
RT extraction
mining.....

pillars
BT support
RT columns
ribs

pipe jacking
BT soft ground tunnelling
RT buried pipes
jacks

pipes
NT buried pipes

piping
failure of an earth mass by erosion and channel formation

plagioclase
BT minerals
RT anorthosite
feldspar
silicates

plane shear failure
BT shear failure

plane strain
RT antiplane strain

planning

plaster
BT structural materials
RT cement
mortar

plastic waves
BT inelastic waves
RT shock waves

plasticity
BT mechanical properties
RT elasticity
inelasticity
liquid limit
strain hardening
strain softening

plastics
RT elastomers
polymers
resins
rubbers

plate bearing tests
BT field tests
RT deformability
strength.....

plate tectonics
BT tectonics
RT earth crust

plates
RT beams
slabs

plexiglas
USE plastics

point load tests
BT lab tests
RT anisotropy
Brazilian tests

Poissons ratio
BT mechanical properties
RT elasticity

polarisation
RT waves.....

polariscopes

pollution
RT environmental effects
environments

polyaxial
SEE triaxial

polymers
RT elastomers
plastics
rubbers

pore fluid pressure
USE pore pressure

pore fluids
UF interstitial fluids
pore water

pore pressure
UF pore fluid pressure
pore water pressure
BT pressure
RT water pressure

pore water
USE pore fluids

pore water pressure
USE pore pressure

pores

porosimetry
USE porosity
AND measurement

porosity
RT capillarity
hydraulic properties.....
permeability
saturation
water content

porous materials

porphyry
BT igneous rocks

position
USE location

post failure behaviour
RT failure.....

postsplitting
USE controlled blasting

potash
BT evaporitic rocks
minerals
sedimentary rocks
RT rock salt

powders

power plants
RT surface structures.....
underground structures.....

powered supports
USE hydraulic props

Precambrian rocks
RT age
Cainozoic rocks
Mesozoic rocks
Palaeozoic rocks

Precambrian soils
RT age
Cainozoic soils
Mesozoic soils
Palaeozoic soils

precast concrete
BT concrete

precipitation
USE rain
OR snow

prediction

preloading
BT loading

presplitting
BT blasting
RT controlled blasting

pressure
BT environments
NT confining pressure
earth pressure
high pressure
hydraulic pressure
hydrostatic pressure
joint water pressure
pore pressure
water pressure
RT suction

pressure cells
RT hydrostatic pressure chambers
testing machines.....

pressure tests
BT field tests
RT hydrostatic pressure chambers

pressure tunnels
BT tunnels

pressure waves
USE P waves

pressuremeters
devices exerting pressure on the sides of a borehole to measure the deformability of the rock
BT instruments

prestressing

prevention

primary waves
USE P waves

principal stress
RT isobars
isoclinics
isostatics
slip lines

probability
RT statistical analysis.....

production

projectiles

propagation
UF transmission (waves)
RT absorption
attenuation
damping
diffraction
dispersion
polarisation
reflection
refraction
scattering
spectra
velocity
waves.....

props
UF supports
BT support
NT hydraulic props
yielding props
RT packing.....

pullout tests
BT field tests
lab tests
RT bolts.....

pulses
UF impulses
RT energy.....

pumping

pumping tests
BT field tests
permeability tests
RT boreholes
permeability

punch tests
BT lab tests
RT bearing capacity
indentation tests
shear strength
tensile strength

pyrite

quarries
RT mines.....
open pit mines

quartz
BT minerals
RT chert
silicates

quartz diorite
USE tonalite

quartz monzonite
BT igneous rocks
intermediate rocks
intrusive rocks

quartzites
BT metamorphic rocks
sedimentary rocks

Quaternary
SEE Cainozoic

quick clays
USE clays
AND sensitive soils

radar
RT electromagnetic methods.....
electromagnetic sounding

radioactive waste
UF nuclear waste
BT waste materials
RT repositories

radioactivity
RT tracers

radiographic tests
BT nondestructive tests

raft foundations
BT foundations

rafts
RT foundations.....

railroads
USE railways

railways
UF metros
underground railways
BT surface structures

rain
RT environments
water
weathering

raise boring
BT boring
RT mining.....
shafts.....

raises
USE shafts

raising
RT mining.....
shafts.....

Rayleigh waves
BT elastic waves
seismic waves
surface waves
RT Love waves

reaming
RT mining.....
shafts.....

rebars
USE bolts

rebound tests
RT Schmidt hammer tests

rectangular
RT geometry

reflection
RT propagation
waves.....

refraction
RT propagation
waves.....

regression analysis
RT mathematical analysis.....
statistical analysis.....

reinforced earth
BT structural materials
RT earth fill

reinforcement
RT support.....

relative permeability
USE permeability

relaxation
BT time dependent behaviour

reliability
RT performance

remedial measures
UF repair

remote control
USE control

remote sensing
UF satellite photography

repair
USE remedial measures

reports

repositories
BT underground structures
RT radioactive waste

research

reservoirs
BT hydraulic structures
surface structures
RT dams.....

residual friction
BT friction

residual soils
NT laterites
RT collapsible soils

residual strain

residual strength
UF ultimate strength
BT mechanical properties
RT critical state

residual stress

resin grouted bolts
BT bolts
grouted bolts
support

resin models
RT model tests

resins
RT grouted bolts.....
plastics
rubbers

resistivity
BT electrical properties
RT conductivity

resistivity methods
BT electrical methods
RT electromagnetic methods.....
geophysics.....
mapping
sounding.....

resonant column tests
BT dynamic tests
lab tests

response
RT performance

retaining walls
BT support
surface structures
walls
RT sheet piles

reviews

rheologic models

rheology
BT mechanics
RT viscosity

rhyolite
BT acidic rocks
extrusive rocks
igneous rocks

ribs
BT support
RT pillars
steel

rigid inclusions
BT inclusions
RT microstructure

rigidity
UF stiffness
BT mechanical properties
RT deformability
shear modulus

ring shear tests
BT lab tests
shear tests
RT direct shear tests.....
residual strength

rings

rippability
BT mechanical properties

ripping
RT rock breaking

rivers
RT geomorphology
hydrology
valleys

road stone
BT structural materials

roadheaders
USE tunnelling machines

roads
UF highways
BT surface structures
RT foundations.....
pavements
railways

roadways (underground)
USE drifts

rock bolts
USE bolts

rock breaking
RT blasting.....
comminution
fragmentation
ripping
rock cutting
splitting

rock burst
RT gas burst
ground movement
rock falls
rock slides

rock classification
BT classification
RT rock mass classification

rock cutting
RT rock breaking

rock falls
RT rock burst
rock slides

rock fill
BT fills
RT gabions

rock fill dams
BT dams
embankment dams
RT dam foundations
gravity dams

rock like materials
RT model materials
structural materials.....

rock mass classification
UF geotechnical classification
BT classification
RT rock classification

rock masses
NT graben
jointed rocks
lenticular orebodies
massive orebodies
massive rocks
orebodies
seams
tabular orebodies
RT soil masses

rock mechanics
use for general material only or to distinguish from soil mechanics
BT geomechanics
mechanics

rock movement
USE ground movement

rock noise
BT noise
RT acoustic emission

rock passes
UF ore passes
waste passes
RT drifts
shafts.....

rock quality designation
BT mechanical properties
RT index tests

rock salt
UF halite
BT evaporitic rocks
sedimentary rocks
RT potash

rock slides
RT avalanches
landslides
rock burst
rock falls
slide mechanisms
slip

rock slopes
BT slopes
RT natural slopes

rock sockets
RT anchorages

rocks
NT acidic rocks
andesite
anhydrite
anthracite
arenaceous rocks
argillaceous rocks
argillites
basalt
basic rocks
breccia
Cainozoic rocks
carbonaceous rocks
carbonate rocks
chalk
chert
clastic rocks
claystones
coal
coal measures
conglomerate
crystalline rocks
diabase
diorite
dolomite
dunite
evaporitic rocks
extrusive rocks
flysch
frozen rocks
gabbro
gneiss
granite
granodiorite
greywacke
gritstones
gypsum
hard rocks
igneous rocks
ignimbrite

rocks (contd)
NT intermediate rocks
intrusive rocks
lava
lignite
limestone
lunar rocks
magma
marble
marls
marlstone
Mesozoic rocks
metamorphic rocks
molten rocks
mudstone
norite
oil shale
Palaeozoic rocks
porphyry
potash
Precambrian rocks
quartz monzonite
quartzites
rhyolite
rock salt
sandstones
saturated rocks
schist
sedimentary rocks
serpentinite
shales
silicate rocks
siltstones
slates
soft rocks
taconite
tar sands
tonalite
tuff
unsaturated rocks
RT aggregates
bedrock
debris
minerals.....
overburden
rock fill
rock masses.....
soils.....

rod mills

roof bolts
USE bolts

roofs
RT underground structures.....

room and pillar
BT mining
RT bord and pillar

rotary drilling

rotational failure
BT shear failure

roughness
BT surface properties
RT abrasiveness
asperities

rubbers
RT elastomers
plastics
resins

rupture modulus
BT mechanical properties
RT bending tests
stress strain relations

S waves
UF distortion waves
secondary waves
shear waves
transverse waves
BT body waves
elastic waves
seismic waves
RT P waves

safety
RT accidents
hazards

safety factors
ratio of resisting moment to failure moment, or, shear strength to shear stress. For general matters use safety
RT slope stability

sag
of a roof, for example, if improperly supported
RT convergence

saliferous rocks
USE evaporitic rocks
OR rock salt

salinity
BT chemical properties

salt mines
RT underground mines

salts
(chemical compounds)
PREFER rock salt (if appropriate)

sampling
RT integral sampling
specimen preparation

sand fill
BT fills

sands
BT soils
NT tar sands
RT aeolian soils.....
aggregates
arenaceous rocks
cohesionless soils
gravel

sandstones
BT clastic rocks
sedimentary rocks
NT greywacke
tar sands
RT arenaceous rocks

satellite photography
USE remote sensing

saturated rocks
RT unsaturated rocks

saturated soils
RT unsaturated soils

saturation
PREFER saturated rocks
OR saturated soils
OR unsaturated rocks
OR unsaturated soils (if appropriate)
RT capillarity
permeability
porosity
water content

scale (mathematics)
USE similitude
OR size effect

scale models
RT model tests

scaling
removing loose rocks from a wall or face
RT mining.....
underground structures.....

scattering
RT propagation
waves.....

schist
BT metamorphic rocks

schistosity
RT cleavage
fissility
foliation

Schmidt hammer tests
RT index tests
rebound tests

screw plate tests
BT field tests

sea bed
RT geomorphology

seams
BT rock masses
RT mines.....
orebodies.....
underground mines

secondary waves
USE S waves

sedimentary rocks
UF mudrocks
NT argillites
anthracite
breccia
carbonaceous rocks
chalk
chert
clastic rocks
claystones
coal
coal measures
conglomerate
dolomite
evaporitic rocks
flysch

sedimentary rocks (contd)
NT greywacke
gritstones
gypsum
lignite
limestone
marls
marlstone
mudstone
oil shale
potash
quartzites
rock salt
sandstones
shales
siltstones
taconite
tar sands
RT carbonate rocks.....
sediments
silicate rocks

sedimentology

sediments
BT soils
RT alluvium
clays.....
gravel
loess
sedimentary rocks.....
silts

seepage
RT drainage
hydrology
leakage
percolation
water flow.....

seismic methods
BT field tests
NT crosshole seismic method
RT geophysics.....

seismic velocity
USE seismic waves
AND velocity

seismic waves
UF strain waves
BT elastic waves
NT body waves
Love waves
P waves
Rayleigh waves
S waves
surface waves

seismic waves (contd)
RT acoustic waves
inelastic waves.....
stress waves
ultrasonic waves

seismicity
the property of exhibiting seismic phenomena, or the degree to which seismic phenomena are present
RT earthquakes

seismographs
USE seismometers

seismology
study of seismic phenomena
BT geophysics

seismometers
BT instruments
RT monitoring systems

seminars
USE conferences

sensitive soils
UF quick clays
RT clays.....
cohesive soils.....
thixotropy

sensitivity
not a soil property - see sensitive soils
RT accuracy
performance

serpentinite
BT basic rocks
igneous rocks
metamorphic rocks

settlement
UF slump
RT consolidation
consolidation tests
oedometers
subsidence
time dependent behaviour.....

shaft linings
BT linings
RT support.....

shaft location
BT location

shaft sinking
RT mining.....

shafts
piles are sometimes referred to as shafts (USA usage)
UF raises
winzes
RT boreholes
mines.....
raise boring
raising
reaming
rock passes
shaft sinking
support.....
underground mines

shales
BT clastic rocks
sedimentary rocks
NT oil shale
RT aggregates
flysch

shallowness
USE depth

shape
RT geometry

shear
PREFER a more specific term (if appropriate)

shear cracks
USE shear fractures

shear failure
NT plane shear failure
rotational failure
RT slip
toppling failure

shear fractures
BT fractures

shear modulus
UF modulus of rigidity
BT mechanical properties
RT stress strain relations

shear resistance
USE shear strength

shear strain

shear strength
BT mechanical properties
NT joint shear strength
RT punch tests

shear stress
RT isochromatics
normal stress.....
slip lines

shear tests
BT field tests
lab tests
NT borehole shear tests
consolidated drained triaxial tests
consolidated undrained triaxial tests
direct shear tests
drained direct shear tests
drained shear tests
drained triaxial tests
ring shear tests
simple shear tests
triaxial tests
unconsolidated undrained triaxial tests
undrained direct shear tests
undrained shear tests
undrained triaxial tests
vane shear tests

shear waves
USE S waves

shear zones

sheet piles
BT support
RT piles
retaining walls

sheeting
USE joint sets

shield supports
USE hydraulic props

shields
tunnelling shields (not a variety of props)
BT tunnelling machines
NT bentonite shield
RT support.....
tunnels.....

shock effects
RT dynamic properties
earthquakes
explosions

shock waves
UF detonation waves
BT inelastic waves
RT plastic waves

shortwall mining
BT mining
RT coal mining

shotcrete
BT support
RT cementation
concrete.....
gunite

shrinkage
RT collapsible soils
expansion
expansive soils
volume
volume strain

silicate rocks

silicates
RT anorthosite
feldspar
mica
minerals.....
muscovite
plagioclase
quartz

silts
RT aeolian soils.....
cohesive soils.....
sediments

siltstones
BT clastic rocks
sedimentary rocks
RT sandstones
shales.....

Silurian
SEE Palaeozoic

similitude
RT size effect

simple shear tests
BT lab tests
shear tests

simulation
BT mathematical analysis
RT mathematical models

site investigation
RT field tests.....

size

size distribution
RT grain size

size effect
RT grain size
similitude

slabs
RT beams
plates

slake durability tests
BT lab tests
slaking tests

slaking

slaking tests
BT lab tests
NT slake durability tests

slates
BT metamorphic rocks
RT aggregates

slenderness
USE thickness

slice analysis
BT mathematical analysis

slickenside
RT faults.....
slip
surface properties.....

slide mechanisms
RT landslides
rock slides
slip

slides
USE landslides
OR slide mechanisms
OR slip

slimes
BT soils
RT tailings dams

slip
RT faults.....
flexural slip
landslides
shear failure.....
slickenside
slide mechanisms
slip surface

slip lines
lines of maximum shear stress
RT graphical analysis
principal stress
shear stress
stress field

slip surface

slope analysis
PREFER slope stability (if appropriate)
BT mathematical analysis

slope angle
RT slope stability

slope stability
BT stability
RT safety factors
slope angle
stability analysis

slopes
PREFER a more specific term (if appropriate)
BT surface structures
NT natural slopes
rock slopes
RT berms
dumps
embankments
open pit mines

slots
RT holes

slump
USE settlement

slump-prone soils
USE collapsible soils

slurries

slurry explosives

smooth blasting
USE controlled blasting

snow
RT ice
permafrost

soft ground
USE soft rocks
OR soft soils (or both)

soft ground tunnelling
BT tunnelling
NT pipe jacking
RT hard rock tunnelling
New Austrian Tunnelling Method
shields.....
tunnelling machines.....

soft rocks

soft soils

soil classification
BT classification

soil compaction
NT vibroflotation

soil formation

soil masses
UF earth masses
RT rock masses.....

soil mechanics
use for general material only, or to distinguish from rock mechanics

soil movement
USE ground movement

soil stabilisation

soil structure interaction
NT pile friction

soils
NT aeolian soils
alluvium
bentonite
Cainozoic soils
calcareous soils
clays
cohesionless soils

soils (contd)
NT cohesive soils
collapsible soils
expansive soils
frozen soils
glacial till
gravel
laterites
loams
loess
lunar soils
marls
Mesozoic soils
mud
organic soils
Palaeozoic soils
peat
permafrost
Precambrian soils
residual soils
sands
saturated soils
sediments
sensitive soils
silts
slimes
soft soils
tar sands
unsaturated soils
RT earth fill
muskeg
overburden
rocks.....
soil masses

solids flow

solifluction
UF mudflows
RT time dependent behaviour.....

solution cavities
BT cavities
RT natural cavities

sonic waves
USE acoustic waves

sonics
USE acoustics

sound emission
USE acoustic emission
OR rock noise

sound waves
USE acoustic waves

sounding
BT field tests
NT electromagnetic sounding
RT electrical methods.....
geophysics
resistivity methods

spalling

specific energy
RT comminution

specimen preparation
RT sampling

spectra

spheres
RT geometry

spiral
RT geometry
helical

split cylinder tests
USE Brazilian tests

split set anchors
BT bolts
support

splitting
RT rock breaking

spoil heaps
USE coal spoil
AND dumps
OR waste materials
AND dumps

springs
structural elements
RT foundations.....

squeezing ground
RT convergence

stabilisation
PREFER soil stabilisation
(if appropriate)

stability
NT slope stability

stability analysis
PREFER slope stability (if appropriate)
BT mathematical analysis

standards
RT contracts
management

statics
BT mechanics

statistical analysis
BT mathematical analysis
NT cluster analysis
factor analysis
Fourier analysis
Monte Carlo method
multivariate analysis
regression analysis
RT probability
stochastic processes

statistics
collections of statistical data
PREFER statistical analysis (if appropriate)

steam
RT environments

steel
BT metals
RT arches
ribs

steepness
USE inclination

stereographic projection
RT graphical analysis

stick slip
RT friction.....

stiff testing machines
BT testing machines

stiffness
USE rigidity

stochastic processes
RT statistical analysis.....

stone
USE masonry
OR rocks

stoping
BT mining
NT open stoping

storage
property of an aquifer, for example
RT hydraulic properties

storage caverns

stowing
USE filling

strain
PREFER a more specific term (if appropriate)
NT antiplane strain
compression strain
finite strain
plane strain
residual strain
shear strain
tensile strain
uniaxial strain
volume strain
RT deformation

strain analysis
BT mathematical analysis

strain energy

strain gauges
BT instruments
RT monitoring systems
strain rosettes

strain hardening
UF work hardening
RT plasticity
strain softening
time dependent behaviour.....

strain measurement

strain meters
USE strain gauges
OR strain rosettes

strain rate
RT time dependent behaviour.....

strain relief
RT doorstoppers
overcoring
stress relief

strain rosettes
RT strain gauges

strain softening
RT plasticity
strain hardening
strength reduction
time dependent behaviour.....

strain waves
USE seismic waves

strata
RT bedding
bedding planes
foliation
layered systems
layers

strata control
PREFER support (if appropriate)

strata movement
PREFER ground movement (if appropriate)

stratification

stratigraphy

strength
BT mechanical properties
NT compressive strength
residual strength
shear strength
tensile strength
triaxial strength
uniaxial strength

strength reduction
UF weakening
RT mechanical properties.....
strain softening

stress
PREFER a narrower term (if appropriate)
NT biaxial stress
compressive stress
effective stress
ground stress
hydrostatic stress
in situ stress
normal stress

stress (contd)
NT principal stress
residual stress
shear stress
tectonic stress
tensile stress
thermal stress
total stress
triaxial stress
uniaxial stress
RT strain.....
stress strain relations

stress analysis
BT mathematical analysis

stress concentration

stress distribution

stress field
RT isobars
isochromatics
isoclinics
isopachs
isostatics
slip lines

stress gradient

stress history
USE stress path

stress measurement

stress meters
BT instruments
RT monitoring systems

stress path
UF stress history

stress ratio

stress relief
UF destressing
RT doorstoppers
overcoring
strain relief

stress state
USE stress
OR in situ stress
OR stress distribution
OR stress field

stress strain relations
UF constitutive models (or laws)
RT bulk modulus
compressibility
deformability
elastic properties
hysteresis
rupture modulus
shear modulus
Youngs modulus

stress waves
RT seismic waves

strike
RT dip
orientation

strike slip faults
BT faults

stripping

structural analysis
BT mathematical analysis

structural geology
RT microstructure

structural materials
NT aggregates
alloys
backfill
bitumen
cement
cemented fill
concrete
earth fill
fills
fly ash
hydraulic fill
metals
mortar
plaster
precast concrete
reinforced earth
rock fill
sand fill
steel
wire mesh
RT masonry
overburden
rock like materials

structure
for rock/soil structure see grain size, microstructure, structural geology

structures
PREFER a more specific term (if appropriate)
NT aerial cableways
arch dams
berms
bridges
buttress dams
caverns
channels
cofferdams
dams
diaphragm walls
drifts
dumps
earth dams
embankment dams
embankments
gravity dams
harbours
hydraulic structures
locks
mines
natural slopes
offshore structures
open pit mines
pavements
piers
power plants
quarries
railways
reservoirs
retaining walls
roads
rock fill dams
rock slopes
shafts
slopes
storage caverns
surface structures
tailings dams
tanks
tunnels
underground mines
underground structures
RT abutments
beams
blocks
faces
floors
roofs
walls.....

subgrades
RT base courses
pavements

subsidence
RT angle of draw
settlement
time dependent behaviour.....

suction
RT pressure.....

sulphides
RT minerals.....

superconductivity
RT conductivity
cryogenic temperature
electricity.....
electromagnetism

support
PREFER a narrower term, or a number of narrower terms (if appropriate)
NT anchors
anhydrite packing
arches
bolts
chemical grouting
grouted bolts
grouting
gunite
hydraulic props
packing
pillars
props
resin grouted bolts
retaining walls
ribs
sheet piles
shotcrete
split set anchors
trusses
wire mesh
wire ropes
yielding bolts
yielding props
RT cementation
fills.....
linings.....
reinforcement

supports
USE props

surface active agents
USE surfactants

surface energy
BT surface properties
RT fracture propagation
energy release rate
surfactants

surface properties
NT abrasiveness
friction angle
roughness
surface energy
RT abrasion
adhesion
adsorption
asperities
friction
hardness
slickenside
surfactants

surface structures
UF buildings
NT aerial cableways
arch dams
berms
bridges
buttress dams
cofferdams
dams
diaphragm walls
dumps
earth dams
embankment dams
embankments
gravity dams
natural slopes
open pit mines
pavements
piers
power plants
quarries
railways
retaining walls
roads
rock fill dams
rock slopes
slopes
tailings dams
tanks
trenches
RT abutments
craters
cuts
foundations.....
hydraulic structures.....
offshore structures

surface waves
BT elastic waves
seismic waves
NT Love waves
Rayleigh waves

surfaces
RT adsorption

surfactants
UF surface active agents
RT surface energy
surface properties.....

surveying
UF levelling
NT airborne surveying
RT distance measurement
mapping

swelling
RT consolidation
expansive soils

sylvinite
BT minerals
ores

symmetry
RT mathematical analysis.....

symposia
USE conferences

tabular orebodies
BT orebodies

taconite
BT sedimentary rocks

tailings dams
BT dams
embankment dams
RT dam foundations
gravity dams

tanks
BT surface structures

tar sands
BT clastic rocks
sands
sandstones
sedimentary rocks
soils
RT oil shale

tectonic stress

tectonics
NT plate tectonics
RT earth crust
geophysics.....

telemetry

telluric methods
BT electrical methods
RT magnetotelluric methods

temperature
BT environments
NT cryogenic temperature
high temperature
low temperature
RT freezing
frost
frozen ground
heat

tensile cut off
RT mathematical analysis.....

tensile fractures
BT fractures

tensile strain

tensile strength
BT mechanical properties
RT bending tests
Brazilian tests
extension tests
hollow cylinder tests
punch tests

tensile stress
BT normal stress

tension

tension tests
BT lab tests
NT uniaxial tension tests
RT beam tests
Brazilian tests

Tertiary
SEE Cainozoic

Terzaghi theory

testing
USE field tests
OR lab tests (or both)

testing machines
NT hydrostatic pressure chambers
stiff testing machines
RT deformation cells
hydraulic cells
pressure cells
triaxial cells

tests
NT field tests
lab tests

textbooks

textiles
includes non-woven materials
UF fabrics

texture
RT microstructure

thaw
RT water

theoretical studies
USE mathematical analysis

theories
USE mathematical analysis
OR one of the following specific theories:
arch theory
Biot theory
catastrophe theory
Darcys law
Griffith crack theory
Mohr theory
perturbation theory
Terzaghi theory
Weibull theory

thermal analysis
RT chemical analysis

thermal conductivity
BT thermal properties

thermal diffusivity
BT thermal properties

thermal drilling

thermal energy
NT geothermal energy

thermal properties
BT physical properties
NT thermal conductivity
thermal diffusivity

thermal rock breaking
USE rock breaking
AND thermal energy

thermal stress

thermodynamics
RT heat
mechanics.....

thermoluminescence

theses

thickness
UF slenderness

thin sections
RT lab tests.....

thixotropy
BT hydraulic properties
RT sensitive soils

three dimensional analysis
BT mathematical analysis
RT dimensional analysis
one dimensional analysis
two dimensional analysis

thrust

thrust faults
BT faults

tides
RT oceans

till
USE glacial till

tiltmeters
BT instruments
RT inclinometers
monitoring systems

timber
USE wood

time dependent behaviour
NT creep
fatigue
relaxation
RT bending tests
creep tests
cyclic loading tests
heave.....
loading rate
settlement

time dependent behaviour (contd)
RT solifluction
strain hardening
strain rate
strain softening
subsidence
uplift
viscoelasticity
viscoplasticity

tonalite
UF quartz diorite
BT igneous rocks
intermediate rocks
intrusive rocks

topography
RT geomorphology

toppling failure
RT shear failure.....

torque

torsion

torsion tests
BT lab tests
RT hollow cylinder tests

total stress
RT effective stress
pore pressure

tracers
RT radioactivity

trackless mining
BT mining

transducers
BT instruments
RT monitoring systems

transmission (waves)
USE propagation

transport (eg moisture transport)
USE diffusion

transverse waves
USE S waves (if appropriate)

trenches
BT surface structures

Triassic
SEE Mesozoic

triaxial apparatus
USE triaxial cells

triaxial cells
RT testing machines.....

triaxial compression
PREFER triaxial tests (if appropriate)

triaxial compression tests
for tests on soils see triaxial tests
BT compression tests
lab tests
RT triaxial tests.....

triaxial shear tests (soils)
USE triaxial tests

triaxial strength
BT mechanical properties
RT compressive strength

triaxial stress

triaxial tests
triaxial compression tests on soils to determine shear and compressive strength and deformation parameters; for rocks use triaxial compression tests
UF triaxial shear tests
unconfined compression tests (soils)
BT lab tests
shear tests
NT consolidated drained triaxial tests
consolidated undrained triaxial tests
drained triaxial tests
unconsolidated undrained triaxial tests
undrained triaxial tests
RT compression tests.....
membranes
triaxial compression tests

trusses
BT support

tsunami
RT earthquakes

tuff
BT extrusive rocks
igneous rocks
RT volcanoes

tunnel linings
BT linings
RT support.....

tunnelling
NT hard rock tunnelling
New Austrian Tunnelling Method
soft ground tunnelling
RT borability
driving
penetration rate
tunnelling machines.....

tunnelling machines
UF roadheaders
NT bentonite shield
shields
RT boring machines
penetration rate
tunnelling.....

tunnels
PREFER tunnelling (if appropriate)
NT pressure tunnels
RT drifts
shields.....
support.....
underground structures.....

tunnels and tunnelling
USE tunnelling

turbidity
BT hydraulic properties

turbodrilling

two dimensional analysis
BT mathematical analysis
RT dimensional analysis
one dimensional analysis
three dimensional analysis

ultimate strength
USE residual strength

ultrabasic rocks
USE basic rocks

ultrasonic tests
BT nondestructive tests

ultrasonic waves
RT acoustic waves
elastic waves.....
P waves
S waves
seismic waves.....

ultrasonics
BT acoustics

unconsolidated undrained triaxial tests
BT lab tests
shear tests
triaxial tests
undrained shear tests
undrained triaxial tests

undercutting
RT mining.....

underground mines
BT mines
underground structures
RT coal mines
drifts
faces
footwalls
gold mines
hanging walls
open pit mines
salt mines
seams
shafts

underground railways
USE railways

underground structures
includes structure components
NT caverns
drifts
shafts
tunnels
underground mines
RT abutments
buried pipes
excavations
faces
floors
mines.....
openings
power plants
roofs
scaling
support.....

underpinning
RT foundations.....
mining.....

undersea mining
BT mining

undrained direct shear tests
BT direct shear tests
lab tests
shear tests
undrained shear tests

undrained shear tests
BT lab tests
shear tests
NT consolidated undrained triaxial tests
unconsolidated undrained triaxial tests
undrained direct shear tests
undrained triaxial tests

undrained triaxial tests
BT lab tests
shear tests
triaxial tests
undrained shear tests
NT consolidated undrained triaxial tests
unconsolidated undrained triaxial tests

uniaxial compression
PREFER uniaxial tests (if appropriate)

uniaxial compression tests
BT compression tests
lab tests
uniaxial tests

uniaxial strain

uniaxial strength
BT mechanical properties
RT compressive strength

uniaxial stress

uniaxial tension tests
BT lab tests
tension tests
uniaxial tests

uniaxial tests
BT lab tests
NT uniaxial compression tests
uniaxial tension tests

unloading
RT loading.....

unsaturated rocks
RT saturated rocks
saturation

unsaturated soils
RT collapsible soils
saturated soils
saturation

uplift
RT heave.....
time dependent behaviour.....

valleys
RT rivers

vane shear tests
BT field tests
lab tests
shear tests

vegetation

velocity
RT propagation
waves.....

vertical

vibrating wire instruments
BT instruments
RT monitoring systems

vibration
RT noise.....

vibroflotation
BT soil compaction

viscoelasticity
RT elastic properties
time dependent behaviour.....

viscoplasticity
RT plasticity
time dependent behaviour.....

viscosity
BT hydraulic properties
RT rheology

void ratio

voids

volcanic rocks
USE extrusive rocks

volcanoes
RT basalt

volume
RT expansion
shrinkage

volume strain
RT expansion
shrinkage

walls
NT diaphragm walls
retaining walls
RT cut offs
footwalls
hanging walls

waste heaps/dumps
USE dumps
AND waste materials

waste materials
NT coal spoil
fly ash
overburden
radioactive waste

waste passes
USE rock passes

water
RT environments
groundwater
humidity
moisture
pore fluids
rain
steam
thaw

water content
RT expansive soils
porosity
saturation

water flow
NT groundwater flow
RT leakage
seepage

water jets
RT cavitation

water pressure
BT hydraulic pressure
NT joint water pressure
RT pore pressure

water table
RT groundwater
hydrology

wave equation
RT mathematical analysis.....

waves
PREFER a narrower term (if appropriate)
NT acoustic waves
body waves
elastic waves
electromagnetic waves
gamma rays
inelastic waves
infra red
Love waves
microwaves
P waves
plastic waves
Rayleigh waves
S waves
seismic waves
shock waves
stress waves
surface waves
ultrasonic waves
x rays
RT absorption
attenuation
damping
diffraction
dispersion
polarisation
propagation
radar
reflection
refraction
scattering
spectra
velocity

weakening
USE strength reduction

wear
RT abrasion

weathering
BT environmental effects
RT air
wind

wedge failure
BT failure

wedges

Weibull theory

well logging
USE borehole logging

wells
RT drainage

wetting

wharves
USE piers

width

wind
RT air
weathering

winzes
USE shafts

wire mesh
BT structural materials
support

wire ropes
BT support

wood
UF timber
BT structural materials

work hardening
USE strain hardening

working faces
USE faces

X ray analysis
BT lab tests

x rays
BT electromagnetic waves

yield criteria
BT failure criteria
mechanical properties
NT Coulomb criterion

yield surface
representation of a yield criterion as a geometric surface

yielding bolts
BT bolts
support

yielding props
BT props
support

Youngs modulus
UF elastic modulus
BT elastic properties
mechanical properties
RT elasticity
stress strain relations